C000129317

"Dictaphone"

Electronic Genius
of Voice and
Typed Word

C. KING WOODBRIDGE

MEMBER OF THE NEWCOMEN SOCIETY
PRESIDENT
DICTAPHONE CORPORATION,
NEW YORK

THE NEWCOMEN SOCIETY IN NORTH AMERICA

NEW YORK   SAN FRANCISCO   MONTREAL

1952

Copyright, 1952
C. K. WOODBRIDGE

🐝

Permission to abstract is granted
provided proper credit is allowed

🐝

The Newcomen Society, as a body,
is not responsible for opinions
expressed in the following pages

🐝

First Printing: August 1952
Second Printing: August 1952

This Newcomen Address, dealing with the
history of Dictaphone Corporation, was de-
livered at the "1952 Connecticut Luncheon"
of The Newcomen Society of England, held
on Skippers' Dock, in Noank, Connecticut,
U.S.A., when Mr. Woodbridge was the
guest of honor, on August 1, 1952

🐝

 SET UP, PRINTED AND BOUND IN THE UNITED STATES
OF AMERICA FOR THE NEWCOMEN PUBLICATIONS IN
NORTH AMERICA BY PRINCETON UNIVERSITY PRESS

*"*DICTAPHONE*"—Electronic Genius
of Voice and Typed Word—A History
An Address in Connecticut*

AMERICAN NEWCOMEN, *through the years, has had*
happy opportunity to honor numerous corporate enter-
prises, both in the United States of America and in
Canada. The present Newcomen manuscript, dealing
with the history of an internationally-known organiza-
tion, is a colorful record of inventive genius combined
with wide practical application. It is a narrative that
intrigues the imagination and gives a clear picture of
important contributions, both in wartime and in peace.

Well-told, of broad human interest, it is a dramatic

recital of American achievement!

🙞 🙞

"As far back as history can take us, we find human curiosity challenged by the mysteries of sound. But there is no record of useful exploration until 1779, when the Russian, Kratzenstein, built a machine which produced the vowel sounds of the human voice. Nine years later, in Vienna, von Kempelen, a noted maker of automata, produced a complete sentence. In 1850, Faber, another Viennese, produced a still more remarkable speech articulating machine. But these were all literally talking machines, creating sound. None of them recorded sound—nor reproduced it."

—C. KING WOODBRIDGE

INTRODUCTION OF MR. WOODBRIDGE, AT NOANK
ON AUGUST I, 1952, BY T. H. BEARD, VICE-
PRESIDENT, DICTAPHONE CORPORATION, BRIDGE-
PORT; MEMBER OF THE CONNECTICUT COMMIT-
TEE, IN THE NEWCOMEN SOCIETY OF ENGLAND.

My fellow members of Newcomen:

A LITTLE OVER forty-five years ago, there was graduated from Dartmouth a good-looking young man over six feet tall and weighing in the neighborhood of 200 pounds. The son of a Congregational Minister, this young man had bred in him a love of mankind and a desire to know and associate with people. The combination of this desire and a boundless energy naturally directed him towards selling. It is, therefore, not surprising that we next see him getting off a train, hiring a horse and buggy, and starting his round of selling and making friends.

ዩ ዩ

The First World War finds our friend heading the sales organization of the Dictating Machine Division of the Columbia Graphophone Company. At the start of the 1920's, we find him convinced that the dictating machine business should be set up as a separate entity. His ability to make friends and his ability to sell enabled him to convince an investment firm owned by two hard-headed Pennsylvania Dutchmen that it would be good judgment to purchase and set up as a separate enterprise the dictating machine business of the Columbia Graphophone Company. The result of these negotiations was the purchase of the business—lock, stock, and barrel—tools, machine tools, patents, and good-will.

❦ ❦

Dictaphone Corporation was set up as an independent entity on January 1, 1923, with the salesman from Dartmouth as its first president and director. The corporation has been a money-making enterprise from the start. Its first president left after four years and became the head of another concern, but retained his interest in Dictaphone Corporation and shortly afterwards was made again a member of its Board of Directors. As the head of the new concern, he reversed the trend of its business and started it towards success. He did the same with several other companies, but his regard for Dictaphone Corporation continued. He was not only a director but he was elected Chairman of the Executive Committee. When, therefore, in 1948, the president of Dictaphone Corporation retired, on account of illness, what was more logical than that the *first* president should again pick up the guiding reins?

❦ ❦

In 1951, Dictaphone Corporation, and its wholly-owned subsidiaries in Canada and England, had its biggest year in sales and profits. It gives me great pleasure to present to you the man who has made this possible, the Dartmouth graduate with a love for his fellow man and a drive to get things done: C. K. WOODBRIDGE.

❦ ❦

The "Leather Bottle" - Cobham.

My fellow members of Newcomen:

A S FAR BACK as history can take us, we find human curiosity challenged by the mysteries of sound. But there is no record of useful exploration until 1779, when the Russian, Kratzenstein, built a machine which produced the vowel sounds of the human voice. Nine years later, in Vienna, von Kempelen, a noted maker of automata, produced a complete sentence. In 1850, Faber, another Viennese, produced a still more remarkable speech articulating machine. But these were all literally talking machines, creating sound. None of them recorded sound—nor reproduced it.

ꙮ ꙮ

The history of Dictaphone Corporation necessarily starts with the history of the recording and reproduction of sound. Sound has intrigued Man through the ages. In the 18th and 19th Centuries, experimentation to record and reproduce sound was conducted by many men without success.

Sound was first recorded by Leon Scott, in 1859. Scott mounted on a horizontal shaft a drum about 5 inches in diameter and 3 inches long. One end of the shaft was threaded. A crank was placed on the other end. When the crank was turned, the drum

[7]

rotated and the threaded end made the shaft and drum move from right to left or vice-versa. Scott made a funnel-shaped mouth-piece, in the bottom of which he mounted a diaphragm. Cemented to the underside of the diaphragm was a hog's bristle. He mounted the mouthpiece so that the hog's bristle rested against the surface of the drum. He smoked the surface of the drum. When he spoke into the mouthpiece, the sound waves of his voice vibrated the diaphragm and made the hog's bristle move. When the drum was rotated, the hog's bristle traced a wavy line in the smoked surface of the drum. This line was the *first* recording of sound. Scott had no way of reproducing this sound but it is interesting to note that, if we had one of Leon Scott's records today, we could reproduce it using a photo-electric cell and vacuum tube amplification.

❧ ❧

The first man to record and reproduce sound was Thomas A. Edison. In 1877, he built an apparatus similar to that used by Scott. It had these important differences. A groove, the pitch of which matched the thread on the end of the shaft, was cut in the surface of the drum. A sheet of tinfoil was wrapped around the drum. Instead of a hog's bristle, Edison attached a ball to the diaphragm. The mouthpiece was adjusted so that the ball pressed on the surface of the tinfoil over the groove cut in the drum. Since the pitch of the groove on the surface of the drum and the pitch of the screw thread on the shaft were the same, the ball attached to the diaphragm always rested on the tinfoil above the groove, no matter how the crank was turned. This meant that the tinfoil could bend as the ball moved. When the crank was turned and the user spoke into the mouthpiece, sound waves vibrated the diaphragm. The ball made a series of dents in the sheet of tinfoil that were proportional to the movement of the diaphragm as it was vibrated by the sound waves of the voice. After the recording was completed, the mouthpiece was brought back to its original position. Turning the drum moved the dented tinfoil beneath the ball. The ball moved up and down in the indentations of the tin-foil. This moved the diaphragm and the recorded sound was re-produced. As Edison himself said, he considered this a scientific toy and went ahead with the development of the electric light.

In 1876, Alexander Graham Bell invented the telephone. For this invention, the Government of France awarded to him the Volta Prize, a sum of 50,000 francs, or, in those days, $10,000, given in memory of the great pioneer and scientist in electricity, Volta. It is interesting to note, by the way, how many names of these early scientists have been preserved in our electrical terms. We have the volt named after Volta, the ampere, the watt, the henry, the farad, named after Faraday, and others.

ೞ ೞ

With the Volta Prize, Alexander Graham Bell founded the Volta Laboratory Association, in Washington, D.C. He staffed it with his cousin, Chichester Bell, a chemical engineer, and Charles Sumner Tainter, a scientist and instrument maker. He assigned to the Volta Laboratory Association the problem of recording and reproducing sound for use in connection with his invention, the telephone. Tainter, Chichester Bell, and Alexander Graham Bell made a three-man team. Alexander Graham Bell supplied the electrical knowledge, Chichester Bell the chemical, and Charles Sumner Tainter the mechanical.

ೞ ೞ

Charles Sumner Tainter kept a very complete record of his experiments, which he called his "home notes." These home notes were started in the early part of 1881. The first entries show that he and Alexander Graham Bell had completed a great deal of study and research on the problem of recording and reproducing sound previous to 1881. These home notes are very interesting reading. In October of 1881, the reference is made in them to two boxes deposited by the Bells and Tainter in the archives of the Smithsonian Institution, with the stipulation that they be opened only with the permission of any two of the three depositors.

In 1937, one of our organization became curious about these boxes which apparently had something to do with the recording and reproduction of sound. Investigation showed that they were still in the archives of the Smithsonian Institution, where they had been placed 56 years before. Alexander Graham Bell and Chichester Bell were dead, but Charles Sumner Tainter, a man in his late eighties, lived in San Diego, California. Upon inquiry, Mr. Tainter stated that he would like to have these packages opened. It seems that in later years Alexander Graham Bell and Charles Sumner Tainter had a disagreement. Not long before he died, Bell wanted to open the Smithsonian boxes, but Charles Sumner Tainter refused his permission. By 1937, however, he decided that he would like to have them opened. In order to do so, it was necessary to obtain the consent of Alexander Graham Bell's heirs. A visit to Dr. Gilbert Grosvenor, a son-in-law of Bell and president of the National Geographic Society, aroused his interest with Bell's surviving heirs, Mrs. Grosvenor and Mrs. David Fairchild, wife of David Grandison Fairchild, the great botanist and explorer. Mrs. Grosvenor and Mrs. Fairchild gave permission to open the boxes. This permission and that of Charles Sumner Tainter were presented to Dr. Abbott, Secretary of the Smithsonian Institution, and, on October 17, 1937, fifty-six years to the day from the time the boxes were placed in the archives, they were brought out and opened.

❦ ❦

It is interesting to see the difference between material used in the 1880's and at the present time. The boxes were wrapped in paper, in appearance like ordinary wrapping paper of today, but still tough and strong. Charles Sumner Tainter had written on the wrapping that the boxes had been deposited on October 17, 1881, by Alexander Graham Bell, Charles Sumner Tainter, and Chichester Bell and were to be opened only with the approval of two of the three depositors. The boxes, approximately 18-inch cubes, were found when the paper was removed to be sealed in tin. By means of a soldering iron and blow torch, the tin lid was removed and the wooden lid taken off.

❦ ❦

In one of the boxes was the first machine on which sound was commercially recorded and reproduced. This machine, in front of me, is an exact copy of the original now in the Smithsonian Institution. It has many points of resemblance to the original Leon Scott device. The points of difference are what make it the basis for the practical recording and reproduction of sound. The drum, instead of being coated with lamp black, is coated with wax and, from Tainter's home notes, we know that chemically it was fifty percent beeswax and fifty percent paraffin. The diaphragm in the mouthpiece carries not a hog's bristle but a steel knife. By means of clamps and adjusting screws, the mouthpiece can be accurately adjusted and locked in place so that the steel knife or stylus projects into the wax a given distance. When the crank is turned, the stylus cuts a groove in the wax. When words are spoken into the mouthpiece, the diaphragm vibrates, moving the knife up and down and cutting a "hill-and-dale" groove in the surface of the wax.

❧ ❧

On the machine in the Smithsonian Institution Charles Sumner Tainter had recorded a quotation from Act I, Scene V, of "*Hamlet*"; "There are more things in heaven and earth, Horatio, than are dreamt of in your philosophy." This recording is without a doubt the oldest recording of the human voice in existence.

❧ ❧

While it was reproduced with a temporary electrical reproducer on October 17, 1937, the Bell and Tainter method of reproduction was quite different. On the back of the machine, you can see a brass tube with a very small hole in the end. To reproduce the sound, Bell and Tainter blew a stream of air at about a hundred pounds per square inch pressure against the tone groove cut in the wax. As the drum turned, the "hill-and-dale" groove modulated the air jet and the words were distinctly heard.

❧ ❧

The other box contained a device which Bell called the Photophone. By means of this device, Bell transmitted the human voice on a light beam for several hundred yards. This was the first transmission of voice without wires.

❧ ❧

You may wonder why the equipment was deposited in the archives of the Smithsonian Institution. The reason was to establish the date when the idea of recording sound by cutting a groove in wax had been put into practical use. By establishing this date, no one could question the fact that Bell and Tainter had done the job prior to October 17, 1881, even though a patent was not issued until some years later. Actually, the Bells and Tainter continued their work and the patent itself was finally issued on May 6, 1886, giving nearly five additional years of life to the patent.

❦ ❦

This Bell and Tainter patent was the basic one for the phonograph industry. There was a law suit over it and the Courts upheld the validity of the patent. All manufacturers of phonographs had to operate under it until its expiration, in 1903.

❦ ❦

With patent protection assured, the Bells and Tainter formed the Volta Graphophone Company and brought into the enterprise Andrew Devine and John H. White, Congressional reporters, and court reporters James O. Clephane and Edward D. Easton.

❦ ❦

Bell and Tainter built several model machines. Bell used one, and another was given to Mr. Easton for his use. He became very enthusiastic over it. He customarily took his notes in shorthand. When he left the House of Representatives, he went to his rooms and immediately dictated his notes, while still fresh, to this model machine. The time saved and the accurate results aroused great enthusiasm on the part of Easton and other court reporters. The first of these models, owned by Alexander Graham Bell, is before me. You will note that there are many changes from the Smithsonian model—instead of a drum with wax coating, a small wax-coated paper tube is used. The sound is recorded by a stylus attached to the diaphragm as in the original model. The recorder is replaced by a reproducer when the record is transcribed. The sound from the reproducer is led to the person typing through rubber tubing and ear tips like a doctor's stethoscope.

[12]

Great interest was aroused by the business possibilities of the machine. A company was formed, and, in 1888, Charles Sumner Tainter went to Bridgeport, Connecticut, rented space from the Howe Sewing Machine Company, and started the manufacture of machines for the recording and reproduction of sound for business purposes. These first machines were driven by the familiar treadle of the old-fashioned sewing machine.

ᘒ ᘒ

The following year, manufacturing operations were moved into space rented in an organ factory located in the western part of Bridgeport. To us it is interesting that machines for recording and reproducing sound for business purposes have been made on that plot of ground continuously since 1889.

ᘒ ᘒ

The company had many vicissitudes, but eventually became the Columbia Graphophone Company, with Mr. Edward D. Easton, the court reporter, as its president. Associated with him were Devine, White, and Clephane. Mr. Easton continued as president until his death in 1912.

ᘒ ᘒ

In the early days, someone sang to this machine and the amusement phonograph business was on its way. The amusement phonograph and record business soon outstripped in volume the business use of the commercial graphophone. To us it is interesting to remember that some of our biggest concerns, like Sears, Roebuck & Company and Westinghouse, have used this equipment for business purposes continuously since the 1890's.

ᘒ ᘒ

Even in those early days the design of the commercial model differed from that of the amusement phonograph. The small paper cylinder with its coating of wax was replaced by a larger solid wax cylinder. The name *"Dictaphone"* was copyrighted in 1906 and became the distinguishing mark of the machines manufactured for business dictation by the Columbia Graphophone Company.

Your speaker became associated with the Columbia Graphophone Company in charge of the sales, service, and distribution of their dictating machine business, in 1918. By 1921, following the expansion after the First World War, the Columbia Graphophone Company had over-played its hand and the receivers for the company assigned to him the problem of disposing of the dictating machine business. The investment banking firm of Swartwout & Appenzellar was convinced by the speaker of the great possibilities of the enterprise. In January 1923, Dictaphone Corporation was formed with Richard H. Swartwout as chairman of the board, Paul Appenzellar a director, C. K. Woodbridge as president and director, L. C. Stowell, now president of Underwood Corporation, as secretary. Mr. Swartwout continued as chairman until his death in 1938, when he was succeeded by Mr. Appenzellar, who has headed our board ever since. His faith and enthusiasm for Dictaphone know no bounds.

❦ ❦

The new corporation purchased—lock, stock, and barrel—dies, jigs, fixtures, patents, and good-will from the Columbia Graphophone Company.

The new enterprise prospered from its inception. The first year showed a satisfactory profit and that profit has steadily grown with the passing years.

❦ ❦

Steps were taken immediately to improve the product. New models were introduced in 1924. In 1933, these were replaced by models of a more advanced design, and, in 1939, the Cameo models, the highest advance in acoustic recording and reproduction, were put into production.

❦ ❦

In 1925, a request came to us from the Pennsylvania Power & Light Company for a machine that would record telephone instructions. Experiments were started on the electronic recording and reproduction of sound. The first concrete result of these experiments was the development of a device called the *Telecord*, its name being derived from the combination of the words "telephone" and "recording." It was attached to a telephone line and recorded electronically both ends of the telephone conversation. Dictaphone telephone recording machines are still in use for this purpose.

❦ ❦

Similar equipment with one or more microphones for the recording of conferences and for police and fire department work was developed. Dictaphone electronic equipment has been used since the late 1920's for this purpose. This development work served as the foundation for electronic dictating equipment.

❦ ❦

The first electronic dictating machines were produced in 1936. A group of them was placed in the hands of one of our large companies to find out how they would operate in everyday use. Interestingly enough, they operated so well that the company purchased these experimental machines. The Dictaphone electronic dictating machine was placed on the market for general sale in 1939.

❦ ❦

In the meantime, Research and Engineering were experimenting with other types of recording media. The basic design of cutting a helical groove on a cylinder has unquestioned superiority. The cylinder is rotated at constant speed. The recorder travels along the cylinder at constant speed. There is no problem of a change in tone line speed. In considering all recording media, Dictaphone has always come back to the scientific principle of a helical groove on a cylinder or a belt, which is really a flexible cylinder.

Experiments were started in 1937 and 1938, using a plastic belt as a recording medium. The result of this experimental work was the introduction and sale, in 1940, of the belt recording machine. This machine uses a plastic belt, five one-thousandths of an inch thick, twelve inches long, and three and one-half inches in width. Many of these were sold in 1940 to the Civil Aeronautics Authority and to other government agencies. Large numbers of these machines were built for the government during the war years.

❧ ❧

During the Second World War, the manufacture of office equipment was suspended and inventories were ear-marked for Government use. Dictaphone devoted its entire effort to the winning of the war. Having the machine tools, the supervisory force, and the trained personnel to manufacture instruments requiring the highest degree of precision, very logically, Dictaphone was asked by the Government to make fire control apparatus. The remote control system which connects the *40 m/m* anti-aircraft gun with the director was first produced by Dictaphone Corporation. Our experience, tool design, process sheets, and know-how were given freely to other producers as the need for more and more of this equipment became necessary. We made more than half of all the remote control systems produced for the U.S. Army.

❧ ❧

Telescopic sight mounts for tanks, gunners' quadrants, computing sights, which automatically calculate the amount the anti-aircraft gun has to lead the plane in order to score a hit, were produced in quantity by Dictaphone Corporation.

❧ ❧

Electronic recording machines, using plastic belts, were produced for the U.S. Signal Corps and used by them for radio intercept work, for the interrogation of prisoners, and for many other purposes. The net result was the award to Dictaphone Corporation of five Army and Navy *E's*.

❧ ❧

With the release of wartime restrictions, the production of our machines for civilian use was once again our prime objective. The electronic cylinder dictating machine, introduced in 1939, was redesigned using the vastly improved components developed during the war. New models were introduced in late '45 and '46, the most efficient wax cylinder machines ever produced.

The new recording machine with plastic belts, used by the Civil Aeronautics Authority and the Signal Corps during the war, commenced to supplant the Telecord using wax cylinders. When restrictions were removed, many were purchased by fire departments, police departments, doctors, societies, and other groups who had occasion to record meetings, conferences, and the like.

ও ও

Experimental dictating machines using the plastic belt were made in 1941, but the war prevented their development. At the war's end, development work was resumed and dictating machines of this type were tested in our laboratory, tested in the hands of consumers, and finally offered for general sale in 1947. This new machine, the Dictaphone *Time-Master*, with its sensational development, the *Dictabelt*, has shown itself to be the outstanding equipment in the dictating machine field. The Dictaphone *Time-Master* has rapidly supplanted the very excellent electronic cylinder machine. Today, better than ninety percent of our present production is of this new model.

ও ও

The heart of this new model machine is the recording medium, the plastic *Dictabelt*. Dictaphone Corporation produces the *Dictabelt* by the extrusion process, a method which many claimed was impossible.

ও ও

Again, in 1952, Dictaphone is contributing its part to the defense effort. We again are making computing sights for anti-aircraft guns, sub-assemblies for range finders for use in tanks, sub-assemblies for aircraft, and special research and development work for the Government.

ও ও

Dictaphone Corporation is proud of its ancestry and is proud of the record it has made, having accumulated in 1951, sales of $21,-000,000 and profits of $1,230,000, the highest in the history of the corporation.

The industry's place in our economic and social complex needs identification. Our function lies in the field of communications. The very life blood of business, the professions, and Government, is the *communication* of information and ideas.

ψ ψ

This, then, is the history of an industry born only sixty-six years ago and the part played by Dictaphone Corporation. Its life span from inception to its present adolescence is within the lifetime of many of us here. What its maturity may bring, no one knows, but I can assure you that it has a colorful past, a sparkling present, and an intriguing future.

THE END

"Actorum Memores simul affectamus Agenda!"

ψ

THIS NEWCOMEN ADDRESS, *dealing with the history
of Dictaphone Corporation, was delivered at the "1952
Connecticut Luncheon" of The Newcomen Society of
England, held at Noank, Connecticut, U.S.A., on Au-
gust 1, 1952.* MR. WOODBRIDGE, *the guest of honor, was
introduced by* T. H. BEARD, *Vice-President, Dictaphone
Corporation, Bridgeport; Member of the Connecticut
Committee, in American Newcomen. The luncheon was
presided over by* FRAZAR B. WILDE, *President, Con-
necticut General Life Insurance Company, Hartford;
Chairman of the Connecticut Committee, in
The Newcomen Society of England.*

❦   ❦

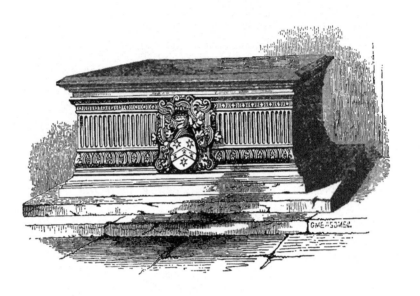

"The history of Dictaphone Corporation necessarily starts with the history of the recording and reproduction of sound. Sound has intrigued Man through the ages. In the 18th and 19th Centuries, experimentation to record and reproduce sound was conducted by many men without success."

—C. KING WOODBRIDGE

"The first machine on which Sound was commercially recorded and reproduced was made by Charles Sumner Tainter and Chichester Bell."

—C. KING WOODBRIDGE

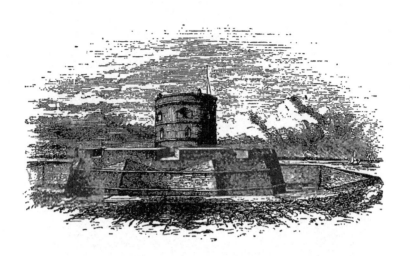

"The first electronic dictating machines were produced in 1936. A group of them was placed in the hands of one of our large companies to find out how they would operate in everyday use. Interestingly enough, they operated so well that the company purchased these experimental machines. The Dictaphone electronic dictating machine was placed on the market for general sale in 1939."

—C. KING WOODBRIDGE

     ツ   ツ

"Our industry's place in the economic and social complex needs identification. Our function lies in the field of communications. The very life blood of business, the professions, and Government, is the *communication* of information and ideas.

"This, then, is the history of an industry born only sixty-six years ago and the part played by Dictaphone Corporation. Its life span from inception to its present adolescence is within the lifetime of many of us here. What its maturity may bring, no one knows, but I can assure you that it has a colorful past, a sparkling present, and an intriguing future."

—C. KING WOODBRIDGE

AMERICAN NEWCOMEN, *interested always in industrial and economic history and in the work and contributions of those whose vision and energy and inventive skill have developed new industries, takes satisfaction in this Newcomen manuscript dealing with the history of a corporate organization—Dictaphone Corporation—whose achievements in the field of electronic recording have been so notable. It is a story typical of the courageous determination and hard work that have made possible the amazing progress of*

American Industry!

ข   ข